NATIONAL
GEOGRAPHIC
KiDS

美 国 国 家 地 理
双 语 阅 读

U0179958

Thomas Edison

托马斯·爱迪生

懿海文化 编著

马鸣 译

第三级

外语教学与研究出版社
FOREIGN LANGUAGE TEACHING AND RESEARCH PRESS
北京 BEIJING

京权图字：01-2021-5130

图书在版编目（CIP）数据

托马斯·爱迪生：英文、汉文／懿海文化编著；马鸣译. —— 北京：外语教学与研究出版社，2021.11（2023.8 重印）
（美国国家地理双语阅读. 第三级）
书名原文：Thomas Edison
ISBN 978-7-5213-3147-9

I. ①托… II. ①懿… ②马… III. ①爱迪生（Edison, Thomas Alva 1847−1931）－传记－少儿读物－英、汉 IV. ①K837.126.1−49

中国版本图书馆 CIP 数据核字 (2021) 第 228167 号

出 版 人　王　芳
策划编辑　许海峰　刘秀玲　姚　璐
责任编辑　姚　璐
责任校对　华　蕾
装帧设计　许　岚
出版发行　外语教学与研究出版社
社　　址　北京市西三环北路 19 号（100089）
网　　址　https://www.fltrp.com
印　　刷　天津海顺印业包装有限公司
开　　本　650×980　1/16
印　　张　37.5
版　　次　2022 年 3 月第 1 版 2023 年 8 月第 4 次印刷
书　　号　ISBN 978-7-5213-3147-9
定　　价　188.00 元（全 15 册）

如有图书采购需求，图书内容或印刷装订等问题，侵权、盗版书籍等线索，请拨打以下电话或关注官方服务号：
客服电话：400 898 7008
官方服务号：微信搜索并关注公众号"外研社官方服务号"
外研社购书网址：https://fltrp.tmall.com

物料号：331470001

Table of Contents

A Great Inventor

Have you ever watched a movie or listened to a recording of music? Have you ever turned on a light? If so, you can thank one man: Thomas Edison! He made all these things possible with his inventions (in-VEN-shuns).

In His Own Words

"Nothing is impossible. We merely don't yet know how to do it."

Thomas Edison worked day and night, with little rest, to make his inventions perfect.

Early Learning

Thomas was born in Milan, Ohio, on February 11, 1847.

The Edison family home in Milan, Ohio

When he was seven, his family moved to Port Huron (HUR-on), Michigan. In school, Thomas daydreamed. His teacher said he could not learn. So, after only three months of school, his mother decided to teach him at home.

Learning from his mother was fun. She taught Thomas to read.

When he was ten, he got a book of science experiments (ek-SPER-uh-ments). He tried every one of them in a lab he built in the family's basement.

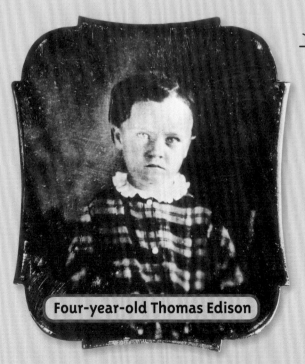

Four-year-old Thomas Edison

Words to Know

EXPERIMENT: A test done to discover or learn about something

LAB: Short for laboratory, a room or building used for science experiments

A Working Boy

Thomas was only 12 when he got his first job. He sold newspapers and snacks to people on a train.

At the train station, Thomas liked to hang out with the telegraph operators. They sent messages to other train stations. Watching them, Thomas knew what he wanted to do next. He would learn to be a telegraph operator.

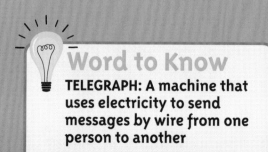

Word to Know

TELEGRAPH: A machine that uses electricity to send messages by wire from one person to another

Young Thomas in his railway uniform

That's a Fact!

When Thomas was working as a newsboy, he built his own telegraph. He and a friend practiced sending messages back and forth from their homes.

In His Time

When Thomas was a boy in the 1850s, many things were different from how they are now.

School

Children often went to school only in winter. They worked on farms or in mines or factories the rest of the time. Many of them did not go to school beyond sixth grade.

Transportation

People traveled by horse and wagon or steamboat. The railroad was just becoming popular.

Communication

The telegraph was the fastest way to send messages. Alexander Graham Bell had not yet invented the telephone.

Toys

Boys played leapfrog and marbles. Girls played hopscotch. Indoors, they all enjoyed the Snake Game. It was a lot like Chutes and Ladders.

U. S. Events

Thomas was 13 years old when Abraham Lincoln was elected president in 1860.

First Inventions

Thomas began work as a telegraph operator when he was 16. He liked to work nights. During the day, he worked on his inventions.

He got his first patent in 1869. It was for a vote-counting machine to be used by legislators (LEJ-is-LAY-turs) in making laws. But no one wanted to buy it.

Thomas's vote-counting machine

Words to Know

PATENT: An official paper that says no one else can make or sell an inventor's work

LEGISLATORS: A group of people who have the power to make and change laws

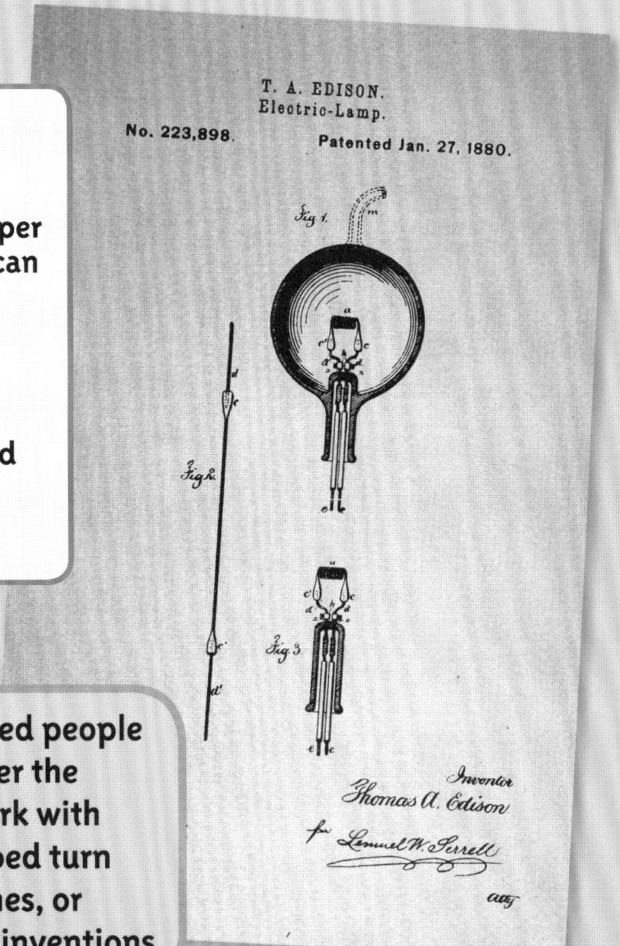

T. A. EDISON.
Electric-Lamp.

No. 223,898.   Patented Jan. 27, 1880.

Fig 1.

Fig 2.

Fig 3.

Inventor
Thomas A. Edison
for Lemuel W. Serrell
atty

That's a Fact! Thomas hired people from all over the world to work with him. They helped turn his ideas and sketches, or drawings, into new inventions.

Thomas did not give up. He worked on other inventions, such as ways to make the telegraph work better.

Always Working

Thomas was so busy working, he did not have time to think about getting married. Then in 1871, when he was 24, he met 16-year-old Mary Stilwell. He liked her right away, and they were married a few months later on Christmas Day.

In His Own Words

"There is no substitute for hard work."

Thomas and Mary had three children. Thomas loved his family, but he did not spend much time with them. He was always working.

In 1876, he opened a lab in Menlo (MEN-loh) Park, New Jersey. There he often worked all day and into the night. When he got tired, he took a nap on a lab table.

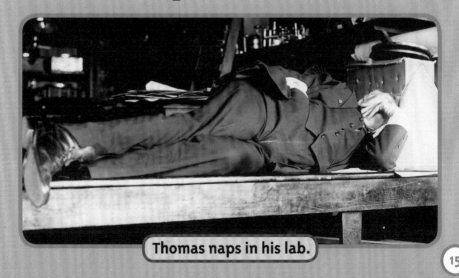

Thomas naps in his lab.

The Wizard

One of Thomas's early inventions was the phonograph, which recorded sound and played it back.

Thomas had an idea for a new machine. Finally, he was ready to test it. He shouted the poem "Mary Had a Little Lamb" into the machine. The machine played it back! Thomas and his workers stayed up all night having fun with the new machine. They called it the phonograph (FOH-nuh-graf).

This invention made Thomas famous. People called him "The Wizard of Menlo Park."

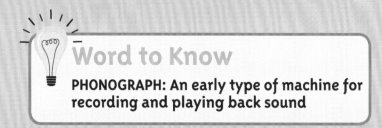

Word to Know

PHONOGRAPH: An early type of machine for recording and playing back sound

7 Awesome Facts

When Thomas was 15, he saved the life of a 3-year-old boy, pulling him from the path of a rolling railroad car.

Thomas nicknamed his first two children "Dot" and "Dash," the short and long sounds used by telegraph operators.

When Thomas and his helpers worked late, they had dinner at midnight. They ate and told stories. Then everyone went back to work.

4

On the train where he sold newspapers as a boy, Thomas also wrote and printed his own newspaper, *The Weekly Herald*. It was the first newspaper printed on a moving train.

5

Of all of Thomas's inventions, the phonograph was his favorite. He called it his "baby."

6

It was Thomas's idea to answer the telephone by saying "hello." Alexander Graham Bell wanted to use the word "ahoy."

7

Thomas and his lab staff filled more than 3,000 notebooks with ideas and sketches for his inventions.

Lights On!

Thomas was soon
working on his next idea.
He wanted to make an
electric light. Electricity
had been around for
a long time, but no
one knew how to light
a house with it.
Instead, people used
candles or oil or gas lamps.

Gas lamp

In His Own Words

"I can never find the thing that does the job best until I find the ones that don't."

One of the filaments that Thomas tested

To make an electric light, Thomas needed something small that could be heated inside a bulb to make it glow. This small thing was called the filament (FIL-uh-ment). Thomas and his workers tried hundreds of things. The best answer was a thin piece of bamboo.

In His Own Words

"Genius is one percent inspiration and ninety-nine percent perspiration."

Thomas had a working lightbulb, but he was not done yet. He made lamps and switches to turn the lamps off and on. He built power stations to make electricity. He used underground wires to carry that power to homes. It took him four years to invent everything he needed to light homes with electricity.

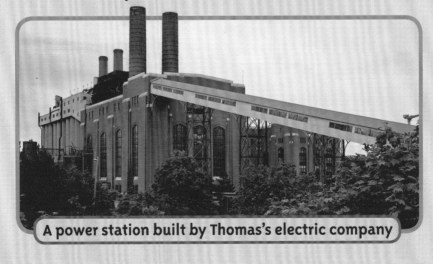

A power station built by Thomas's electric company

Changes

In 1884, Mary Edison died. Thomas was sad and lonely. Then in 1885, he met Mina Miller, and they married a year later.

That's a Fact!

Thomas had six children—a daughter and two sons with his first wife, Mary, and another daughter and two more sons with his second wife, Mina.

Thomas and his wife, Mina, enjoyed an automobile ride.

Inside Thomas's lab

Thomas's new home in New Jersey

They moved into a 23-room house in West Orange, New Jersey. There Thomas built a new lab much larger than the one in Menlo Park.

Always Inventing

Thomas went back to work on his phonograph. He had invented it to record people speaking. Now he began making records so people could listen to music on it. He also invented a movie camera and a machine to show short movies.

One of many styles of phonograph Thomas built. This one is from 1905.

1847	1854	1863	1869	1871
Born on February 11 in Milan, Ohio	Moved to Michigan	Began work as a telegraph operator	Got his first patent	Married Mary Stilwell on December 25

Thomas began losing his hearing when he was a child. He said it helped him be a better inventor because he wasn't distracted by noises around him while he was working.

Thomas's machine for viewing movies

Thomas worked on many other inventions, too. In all, he got 1,093 patents. That was more than any other inventor until the year 2000.

1876	1877	1879	1886	1931
Opened his lab in Menlo Park, New Jersey	Invented the phonograph	Created a bulb that provided hours of light	Married Mina Miller on February 24	Died on October 18

A Great Honor

Thomas began experimenting when he was a boy. More than 70 years later, he was still inventing! He worked until a few months before his death on October 18, 1931. He was 84 years old.

Thomas Edison (right) with President Herbert Hoover

People across the country were sad. At 10 p. m. on October 21, 1931, they turned off their lights for one minute. It was a way to thank the man who had brought electricity into their homes and changed their lives. Thomas Edison made the world a better, easier place to live. We still use his inventions every day.

In His Own Words

"I think work is the world's greatest fun."

Quiz Whiz

See how many Edison questions you can get right!

Answers are at the bottom of page 31.

When Edison was a boy, people _____.

A. Watched television
B. Talked on the telephone
C. Used candles and oil or gas lamps to light their homes
D. Traveled by plane

Before the telephone was invented, the fastest way to send messages was by _____.

A. The U.S. Postal Service
B. Email
C. Telegraph
D. Steamboat

Edison's first patent was for _____.

A. A telegraph
B. A phonograph
C. A movie camera
D. A vote-counting machine

4

Edison had _____ children.
A. Three
B. Four
C. Five
D. Six

5

Edison invented _____.
A. A phonograph
B. Ways to make the telegraph work better
C. A movie camera
D. All of the above

6

Edison got _____ patents.
A. About 3,000
B. More than 1,000
C. About 500
D. More than 2,000

7

People honored Edison a few days after his death by _____.
A. Building a large statue of him
B. Playing music on their phonographs
C. Turning off their lights
D. Flying American flags

Glossary

EXPERIMENT: A test done to discover or learn about something

LAB: Short for laboratory, a room or building used for science experiments

LEGISLATORS: A group of people who have the power to make and change laws

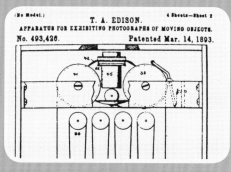

PATENT: An official paper that says no one else can make or sell an inventor's work

PHONOGRAPH: An early type of machine for recording and playing back sound

TELEGRAPH: A machine that uses electricity to send messages by wire from one person to another

第 4—5 页

伟大的发明家

你看过电影，或听过音乐唱片吗？你开过电灯吗？如果是，那你要感谢一个人：托马斯·爱迪生！他用他的发明让这些事情成为可能。

爱迪生语录

"一切皆有可能。我们只是还不知道该怎么做。"

为了完善发明，托马斯·爱迪生夜以继日地工作，很少休息。

第 6—7 页

早期学习

1847 年 2 月 11 日，托马斯在俄亥俄州的米兰镇出生。

在他 7 岁时，他们一家搬到了密歇根州的休伦港。在学校里，托马斯整天幻想。他的老师说他无法学习。因此，仅仅在学校里待了三个月后，他的妈妈就决定在家里教他。

跟着妈妈学习非常有意思。她教托马斯阅读。

在 10 岁时，他得到了一本关于科学实验的书。他在自家的地下室里建了一间实验室，把书上的实验全都做了一遍。

爱迪生在俄亥俄州米兰镇的老家

4岁时的托马斯·爱迪生

小词典

实验：为了发现或者了解某些东西而进行的测试

实验室：英文单词laboratory的简写，指用来做科学实验的房间或建筑物

第 8—9 页

工作的男孩

托马斯找到第一份工作时才 12 岁。他在火车上向人们兜售报纸和零食。

在火车站时，托马斯喜欢和报务员待在一起。他们向别的火车站发送消息。看着他们，托马斯知道自己接下来想要做什么了。他要学习成为一名报务员。

小词典

电报机：一种机器，通过电线用电把信息从一个人这里发送到另一个人那里

穿着铁路制服的小托马斯

不可思议的事实

当托马斯做报童时，他制造了一台属于自己的电报机。他和一个朋友练习从自家来回向对方发信息。

▶ 第 10—11 页

他所在的时代

19 世纪 50 年代时，托马斯还是一个小男孩，很多东西都和它们现在的样子不一样。

学校

孩子们通常只在冬天上学。其余时间他们都在农场上、矿场上或者工厂里干活。很多孩子读完六年级就不上学了。

交通

人们出行乘坐马车或者汽船。铁路才刚刚兴起。

通信

电报是传送信息最快的方式。亚历山大·格雷厄姆·贝尔还没有发明电话。

玩具

男孩玩跳背游戏和弹珠游戏。女孩玩跳房子游戏。在室内，他们都喜欢玩蛇形棋，它很像蛇梯棋。

美国大事件

1860年，亚伯拉罕·林肯当选为总统时，托马斯13岁。

▶ 第 12—13 页

第一批发明

16 岁时，托马斯开始做报务员。他喜欢上夜班。白天的时候，他忙着搞发明。

1869 年，他获得了第一项专利。那是一台供立法者在制定法律时使用的计票机。但是没有人想买下它。

托马斯没有放弃。他致力于别的发明，例如改进电报机的方法。

托马斯的计票机

小词典

专利证书：一种官方文件，宣告他人不得制造或售卖发明者的成果

立法者：有权力制定、修改法律的人

不可思议的事实

托马斯雇用了来自世界各地的人和他一起工作。他们帮忙把他的想法、草图或画稿做成新发明。

▶ 第 14—15 页

总是在工作

爱迪生语录

"勤奋无可替代。"

托马斯忙于工作，以至于他没时间去想结婚这件事。1871 年，当他 24 岁时，他遇到了 16 岁的玛丽·史迪威。他对她一见钟情，短短几个月后，他们就在圣诞节那天结婚了。

托马斯和玛丽生了三个孩子。托马斯很爱家人，但他并没有花太多时间陪伴他们。他总是在工作。

托马斯在他的实验室里打盹儿。

1876 年，他在新泽西州的门罗公园建了一间实验室。他经常在那里没日没夜地工作。如果累了，他就在实验室的工作台上打个盹儿。

▶ 第 16—17 页

"魔法师"

不可思议的事实

托马斯认为他的留声机可以用来为盲人录制书籍。

托马斯有了一个想法，想要制造一台新机器。终于，他准备好测试它了。他冲着机器大声朗读了一首诗：《玛丽有只小羔羊》。机器竟然回放了这首诗！托马斯和工人们用这台新机器玩了一整夜。他们给它取名叫"留声机"。

这项发明让托马斯声名远扬。人们称他为"门罗公园的魔法师"。

托马斯的早期发明之一是留声机，它可以录下声音并回放。

小词典

留声机：早期的一种用于录音和回放的机器

▶ 第 18—19 页

7 个令人赞叹的事实

1 托马斯15岁时，他救了一个3岁的男孩，把他从一辆正在行驶的有轨电车下拉了出来。

2 托马斯给他的头两个孩子取小名叫"嘀"和"嗒"，报务员每天都要使用这一长一短两个声音。

3 如果托马斯和助手工作到很晚，他们会在半夜吃饭。他们一边吃一边讲故事。然后大家继续工作。

4 在那列他在孩童时卖报纸的火车上，托马斯也编写并印刷了他自己的报纸——《通讯周刊》。这是第一份在行驶的火车上印刷的报纸。

5 在托马斯所有的发明中，留声机是他的最爱。他称它为"宝贝"。

6 接电话时说"喂"是托马斯的主意。亚历山大·格雷厄姆·贝尔想用"嘿"这个词。

7 托马斯和他的实验室员工把3000多个笔记本都写满了，上面全是与他的发明相关的想法和草图。

▶ 第 20—21 页

灯亮了！

托马斯很快就投入到他的下一个想法中。他想制造电灯。电已经出现很长时间了，但没人知道怎么用它照亮房间。人们还在用蜡烛、油灯或煤气灯照明。

为了做出电灯，托马斯需要一种可以在灯泡里加热的小东西，以使它发光。这种小东西叫"灯丝"。托马斯和他的员工试了上百种东西。效果最好的是竹丝。

煤气灯

爱迪生语录

"只有看到不完美，我才知道完美应该是怎样的。"

托马斯测试过的一种灯丝

▶ 第 22—23 页

爱迪生语录

"天才是百分之一的灵感加上百分之九十九的汗水。"

托马斯有一个能照明的电灯泡，但是他并没有完工。他制作了电灯和用来打开、关闭电灯的开关。他建造了发电厂来发电。他用地下电缆为家家户户输送电能。他用四年的时间发明了用电照亮每个家庭所需的一切。

托马斯的电力公司建造的一座发电厂

▶ 第 24—25 页

变故

1884 年，玛丽·爱迪生去世了。托马斯既伤心又孤单。1885 年，他认识了麦娜·米勒，一年后他们结婚了。

不可思议的事实

托马斯有六个孩子——和第一任妻子玛丽生了一个女儿、两个儿子，和第二任妻子麦娜也生了一个女儿、两个儿子。

他们搬到了位于新泽西州西奥兰治一栋有 23 个房间的房子里。在这里，托马斯建了一间新的实验室，比门罗公园的那个大很多。

托马斯和他的妻子麦娜乘坐汽车旅行。

托马斯的实验室内部

▶ 第 26—27 页

发明不止

托马斯继续改进他的留声机。之前他发明它是为了录下人们说的话。现在他开始制作唱片，这样人们就可以用它来听音乐。他还发明了一台电影摄像机和一台播放电影短片的机器。

托马斯还致力于很多别的发明。他总共获得了 1,093 项专利。在公元 2000 年之前，那比任何一个发明家都多。

托马斯在新泽西州的新家

托马斯制造的众多留声机中的一种。这台来自于1905年。

托马斯发明的电影放映机

不可思议的事实

幼年时，托马斯的听力就开始衰退。他说这有助于他成为更优秀的发明家，因为他在工作时不会因为身边的噪音而分神。

1847年 于2月11日出生在俄亥俄州的米兰镇	**1854年** 搬到密歇根州	**1863年** 开始做报务员	**1869年** 获得第一项专利	**1871年** 于12月25日和玛丽·史迪威结婚
1876年 在新泽西州的门罗公园建了他的实验室	**1877年** 发明留声机	**1879年** 制造出可以连续数小时发光的灯泡	**1886年** 于2月24日和麦娜·米勒结婚	**1931年** 于10月18日逝世

▶ 第28—29页

崇高的荣耀

还是一个小男孩时，托马斯就开始做实验了。70多年后，他还在搞发明！他在1931年10月18日去世，在那之前的几个月，他依然坚持工作。他享年84岁。

全国人民都非常伤心。1931年10月21日晚上10点，他们熄灯1分钟。人们用这种方式来感谢这个把电送到千家万户、改变了大家生活的男人。托马斯·爱迪生让世界变得更美好、更便捷。现在我们仍然每天都使用他的发明。

托马斯·爱迪生（右）与总统赫伯特·胡佛

爱迪生语录

"我认为这个世界上工作最有趣。"

答题小能手

看看你能答对几个有关爱迪生的问题！答案在第 31 页下方。

当爱迪生还是个小男孩时，人们 _____。
A. 看电视　　　　　B. 用电话聊天
C. 用蜡烛、油灯或者煤气灯为家里照明
D. 乘坐飞机旅行

在电话被发明之前，传送消息最快的方式是 _____。
A. 美国邮政服务　　B. 电子邮件
C. 电报　　　　　　D. 汽船

爱迪生的第一项专利是 _____。
A. 电报机　　　　　B. 留声机
C. 电影摄像机　　　D. 计票机

爱迪生有 _____ 个孩子。
A. 3　　B. 4　　C. 5　　D. 6

爱迪生发明了 _____。
A. 留声机　　　　　B. 改进电报机的方法
C. 电影摄像机　　　D. 以上都是

爱迪生获得了 _____ 项专利。
A. 大约 3,000　　　B. 1,000 多
C. 大约 500　　　　D. 2,000 多

爱迪生去世几天后，人们用 _____ 来纪念他。
A. 为他建造一座大雕像　　B. 用留声机播放音乐
C. 熄灯　　　　　　　　　D. 悬挂美国国旗

词汇表

实验：为了发现或者了解某些东西而进行的测试

实验室：英文单词laboratory的简写，指用来做科学实验的房间或建筑物

立法者：有权力制定、修改法律的人

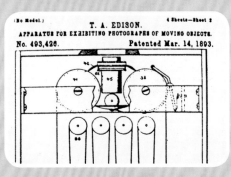

专利证书：一种官方文件，宣告他人不得制造或售卖发明者的成果

留声机：早期的一种用于录音和回放的机器

电报机：一种机器，通过电线用电把信息从一个人这里发送到另一个人那里